A Year at Sea on HMS Implacable 1909

From the 1909 Diary of
Albert 'Ajax' Adams

All proceeds donated to
HMS Ajax & River Plate Veterans Association

Copyright © 2012 Robert Adams

All rights reserved. No part of this publication may be reproduced or transmitted in any form or by any means, electronic or mechanical including photocopying, recording or any information storage or retrieval system, without prior permission in writing from the publishers.

First published in the United Kingdom in 2012 by
Ajax Adams Press

ISBN 978-0-9573729-0-0

All the original handwritten script has been included in type for easier reading. Where considered interesting, the original is included as a graphic. Additional words to make notes easier to read are denoted by *italics*. Technical terms, naval expressions and slang are explained by footnotes. Some illustrations are of Albert's photos and notes from the diary and his students sketches.

**Acknowledgement is made to Science Museum, London for permission to reproduce its photos.
Thanks to Lindy Lovegrove for permission to reproduce images from her collection of original postcards.**

A Year at Sea on HMS Implacable 1909
Albert 'Ajax' Adams

Naval Constructor

Key map

Contents

Preface		vi
Chapter 1	I Get my Bike on Board	1
Chapter 2	Off to Sea at Last	9
Chapter 3	Ireland and Scotland	17
Chapter 4	A New Captain	25
Chapter 5	The Naval Reviews	37
Chapter 6	Battle Practice	57
Chapter 7	Admirals' Inspections	65
Chapter 8	To Gibraltar	75
Addendum		88

Preface

Albert was born 1885, the youngest of 8 children in Pembroke Dock to a ship building family of strong Wesleyan beliefs. Educated at the local school, and then apprenticed as a Shipwright in the Royal Naval Dockyard. After night study, he gained a scholarship to the prestigious Royal Naval College, Greenwich to study Naval Architecture. On graduating, he served time at the Royal Dockyard, Chatham, including the training of boy shipwrights, from where his diary takes up the story.

The Admiralty knew its Fleets were outdated. Admiral Beresford's contribution to solving the problem was to give future Naval Constructors the experience and practicality of the current designs by serving a year-long experience at sea on board ship. Albert was one of the first Assistant Constructors to be posted to the Fleet under Admiral Beresford's programme in 1909.

For his sea time, he was posted to HMS Implacable. This ship had been launched in an incomplete state 1899 in order to clear the dry-dock for construction of a battleship and was never completed properly.

She needed refits in 1902, 1903, 1904, 1905, and again after a boiler explosion in 1906.

The last refit was in Chatham Dockyard, where her appearance changed, as in photos, by building the distinctive Fire Control Platforms on the fore-mast, a wider bridge and extra armour plating.

Admiral orders that Naval Constructors must 'Go to Sea'.

In 1908, Admiral Charles Beresford ordered that Assistant Constructors on graduating from the Royal Naval College, Greenwich, be sent to sea in His Majesty's ships. He stated that he wanted to secure for the future Navy, the best advantage of their training, knowledge and experience. He ordered that Assistant Constructors to be afforded every facility for obtaining a practical acquaintance with the working of the Ship and that these Officers were not to be treated as Watchkeep Officers.
There was no objection to their performing a certain amount of Watchkeeping if found to be instructive to them. However, he laid down that under the responsible officer they had a duty to pay special attention to the practical maintenance of hull and fittings including those of:-
Pumping, Flooding (including magazine) and Draining. Fresh and Salt Water Service. Pumps. Capstan Gear, Steering Gear. Watertight doors, scuttles, hatches and gear for operating them. Ship ventilation including magazine cooling arrangements. Masts, boats and anchors. Cooking apparatus and bread-making plant.
They had to assist in the preparation of Defect Lists of Hull and Fittings and of proposals for alterations and additions: also of periodic reports of surveys of plates and watertight compartments and return of rolling. They were to be detailed for duty with the Ships Staff carrying out repairs to hull and fittings. They had to keep a Journal in which the results of their observations, if opportunity arose, should be noted and which on the completion of their service afloat had to be forwarded to the Admiralty. The Admiral identified some of

these observations such as-

Rolling and pitching of ship under various conditions at sea. Vibration of ship at different speeds. Lengths, heights and period of waves. Manoeuvring capabilities of ship. Effect on steering of different forces and directions of wind. Extent of target offered by ships of the fleet in various conditions of weather. Angles of heel produced by circling and gun firing. Towing ship and being towed, including revolutions, speed and power used in towing. Hoisting boats, mooring and unmooring by hand and power. Laying out anchors.

Rigging net defence and collision mats. Clearing for action. Collision and fire quarters. Embarking ammunition, torpedoes, provisions and stores. Coaling and oil fuelling in harbour and at sea. Fire control.

Effect on speed of different draughts and coatings on bottom, time out of dock. Effect of gun fire and blast on structures and fittings. Housing and striking topmast and topgallant masts carrying aerial wire. Messing, sleeping, berthing, cabin and living arrangements. Lighting of the ship, electric and secondary.

Effect of external fittings on wetness of ship. Communications by navy phones, voice pipes. Accommodation ladders, awnings, side screens and other topside fittings. Magazine and shell room arrangements. Torpedo arrangements. Anchor and cable arrangements. Upper and lower conning tower arrangements. Pumping arrangements. Restoration of trim by flooding compartments if damaged in action. Fresh and salt water services including taking in reserve feed and drinking water from alongside and test tank arrangements, watering boats. Engine and boiler room access and ventilation arrangements. Stowage and handling of oil fuel generally. Masts, spars and rig. Sounding arrangements. Bridge and compass arrangements. Workshop arrangements. Turret aprons. Access to turrets and handing ammunition to guns on roofs. Searchlight arrangements.

Their presence on board could be made use of at the discretion of the Commanding Officer in giving instruction to the Junior Officers in regard to Ships Construction as a means of assisting them in qualifying for their Seamanship.

Admiralty Memorandum N°675 to Channel Fleet 10th Sept 1908 (4959)

Chapter 1   I Get my Bike on Board

Monday 1st February 1909
Last day in H.M. Dockyard Chatham. Transferred Boy Shipwrights over to Scott. Arrangements for removal of *kit* boxes from *my digs* at Harrowdale, 107 Rock Avenue, *Chatham* to Implacable.

Tuesday 2nd February
Took Cab from Rock Avenue to HMS Implacable lying in No. 3 Basin H.M.Dockyard Chatham. Reported to Engineer Commander Crowley on Implacable and then to Captain on HMS Venerable, officers and crew of which were paying off and turned over to Implacable. Reported to Comm^r Carr on Venerable who informed me that my appointment had been cancelled. The shock was ~~only momentary~~ great but discovered that he mistook me for the Paymaster on account of my Shipwrights silver-grey and Paymaster's white bands appearing very much alike[1].

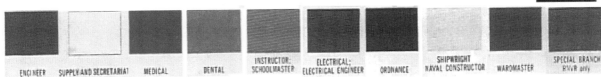

As Assistant Constructor, my official dress uniform is Frock Coat and Sword.
Unpacked tin case and brought my books from the office at the dockyard.
Went ashore in the evening to bring gear from Rock Avenue and left my bicycle at *cousin* Dick Adams'.
Found bunk very strange.

---

[1] These coloured bands between the gold rings on the sleeve cuffs identified the role of non deck officers. Albert's photos on cover and page 63 shows no gold ring curl, as this did not appear for Naval Constructors and Engineers until 1915.

Wednesday 3rd February

Breakfast 8.15. Coaled ship from colliers alongside at 9am till 8.30 in evening. 950 tons lifted by means of stump derrick on Gun and Fcl[2] decks and by using colliers' winches. I was detailed for charge of Double Bottom[3] and Watertight Doors.

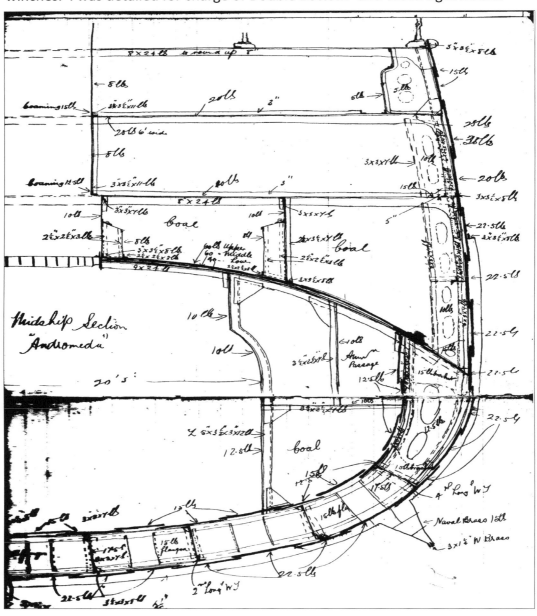

---

[2] forecastle fo'c's'le (fōk'səl) upper deck of a ship at the bow forward of the foremast.

[3] as seen in his own shipwright's notebook sketch of HMS Andromeda.

Went to dockyard for books and papers at office morning and afternoon.
Turned in 11pm.

Thurs 4th Feb

Breakfast 8.30. Coaled ship from 8am till 2pm. 616 tons.
Went to office at dockyard for more books.
Letter from *brother* George in Gibraltar.
Went ashore at 6.30 to class at Wesley Chapel, Arden Street, arriving back at 10pm.
Ordered photographic material at Boots'.

Friday 5th Feb

Turned out to bath 8am and discovered that care was necessary in using sponge bath as otherwise most of water would be outside bath.[4]

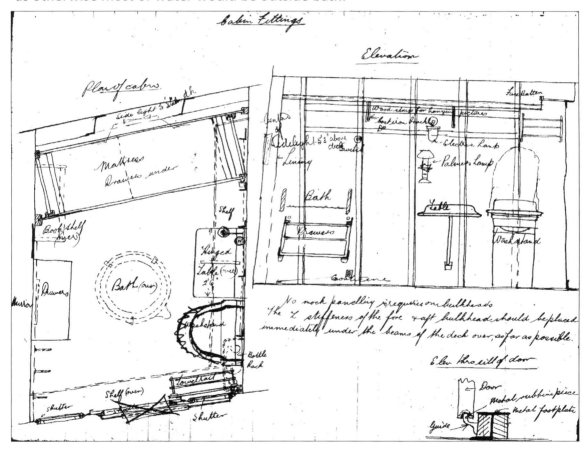

Electric trials by dockyard. Lent my sword to Engr Lt Hamblin for the Chaplain's wedding. Ashore in evening to supper at Dick Adams'.
Brought my bicycle on board.

---

[4] The bath was nothing more than a bowl to stand in as in his sketch of a cabin.

Saturday 6th February

Went to London and had a good look round bookshops in Charing Cross Road. Cinematograph "Hales tour to Messina before and after earthquake".
Took motor bus to Canning Town for meeting. Slept night at 'Sailors Welcome Home'.

Sunday Feb 7th

February Communion Service with West London Mission at Lyceum Theatre, emphasising similarity between communion service and mementoes we have of friends who have passed away and how these keepsakes remind us of lovely incidents in our experience with friends and not of death.
Tram to City Road in evening, left Charing Cross 8.25.
Arrived back at ship 10pm.

Monday 8th February

3 hours Basin Trial after refit.

Taking in ammunition from lighters alongside by means of stump derricks on shelter decks.

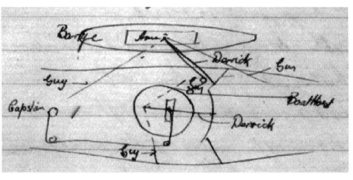

Ashore in evening and bought pictures and *photo* lampshades etc.

Tuesday 9th February

Provisioning and stores on ship. Forwarded subscription to Institution of Naval Architects and Naval Constructors Corps fund.

Called on Dick Adams in evening and borrowed book "The Smoking Craze" by T.Ballard on tobacco.

Wednesday 10th

Ship left *Chatham* Basin at 11am for South Lock and left at 2pm for Sheerness. Moored to buoy at 2.30.

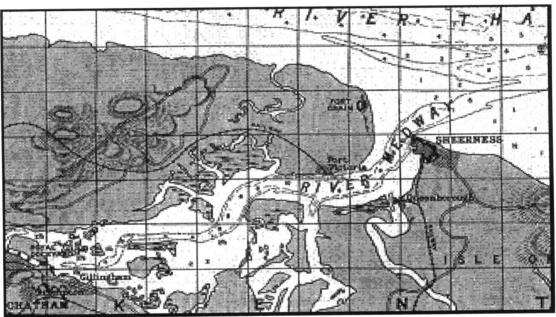

All compartments in double bottoms closed up preparatory to going to sea.
Boats got into harbour positions[5].

---

[5] Boats used as tenders for transport to and from shore

Thursday 11th February
Drawing torpedoes and clinking[6] down.
12"ammunition struck down through Barbette by ordinary hydraulic tray.
8am to 8pm.

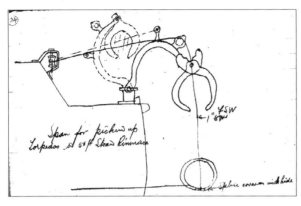
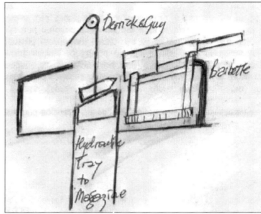

Friday 12th February
Taking more Ammunition as on Thursday.
Examined hinged watertight doors on taking over from Dockyard.

Saturday 13th
Repairing and opening up pipes to draft indicator[7] till 4.30pm.

---

[6] Hoisting an object by a rope tied at centre of gravity so it could be manoeuvred through awkward spaces.
[7] The Indicator was a pressure gauge mounted below the water line with a pipe connected directly to the outside sea water thus continuously reading the depth below waterline. By adding the vertical depth of the bottom of the hull (keel) below the gauge, the draft could be continuously read off in feet if calibrated correctly. This reading would be transmitted to the bridge over voice tube when requested.

Sunday 14th February

Divisions at 9.30. Heads of Departments and Commander Carr & Captain Fawckner inspected the ship.

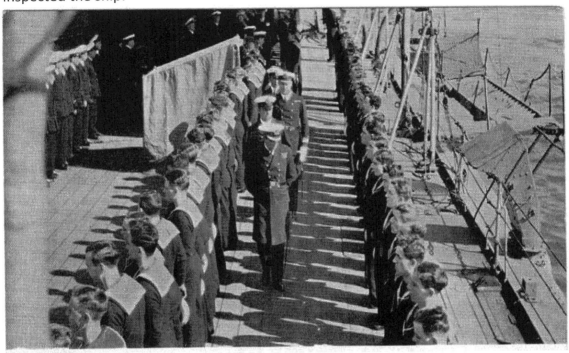

Of course, made the mistake of following the Commander on his inspection with the Heads of Department.
Frock coat and sword.
All messes and deck flats[8].
Afternoon went ashore at Sheerness pier, called at Uncle Tom Adams and went to Bible Class at Congregational Church.
Walked along cliff towards Minster.
Spent evening at 44 Maple Street and went aboard at 10pm after waiting 1 hour at Sheerness pier.

---

[8] Deck flat was an unenclosed clear area.

Monday 15th February
Magazine cooling trials of 12", 6" and 12 pounder guns.
Measured rate of air supply with anemometer and temperature with thermograph in each magazine, prior to actual trials.

Tuesday 16th February
Receipt from Institution of Naval Architects and Royal Corps lapel badge.

Preparing lectures on Ship Construction and Stability for Seamanship.

Wednesday 17th February
Instructing Midshipmen[9] on Stability[10] for their examinations.

---

[9] Trainee Royal Naval officers

[10] The Navy's requirement of mounting heavy guns as high as possible made the Ships stability critical for Naval Architects. It was important that future officers understood and be able to calculate the stowage of the cargo and ballast efficiently.

Chapter 2   Off to Sea at Last

[Handwritten roster of officers, HMS Implacable on re-commissioning Feb 2nd 1909:]

- Captain — Fawckner (left March 31st .09)
- Commander — Barr
- Lieutenants — T 1st Way
- " G Lockyer
- " N Scott
- " Longstaff
- " Cookson ⎱ Watch
- " Coxon ⎰ Keepers
- " P Ridge Jones (left May 09)
- " Gardner
- " RNR Gardner
- Eng Commander — EAB Crowley (left Dec 1910)
- " Lieut 1st — G Robertson
- " — EJ Broker
- " — RC Brown
- Asst Const. — A Adams (left Feb 1910)
- Captain RMLI — Milner
- Lieutenant RMLI — Robinson
- Chaplain + Naval Instructor — McKew (left April 10)

- Fleet Surgeon — Robley Browne (left May 09)
- Staff Surgeon — O'Malley (left June 09)
- Fleet Paymaster — Horniman (left Feb 10)
- Asst Paymaster — Bourne
- Sub Lieut — Gardner (Promoted to Lieut June 09)
- " — Mortimer (Promoted Dec 09)
- " — Taylor (Promoted to Lieut July 09) (left Sept 09)
- Capt — Hughes
- Asst Eng — Barker
- " — Jackson
- Gunner — Wallace
- " — Hawton
- Torp Gunner — Deeney (left Nov 09)
- Boatswain — Davies (left Feb 10)
- " — Langford
- Midshipman Black, Janion (Promoted to Asst Pay), Smiles (Promoted to Asst Pay) (left Sept 09)
- Midshipman Courtney, Strickland, White, Agar

Thursday 18th February
Letter from Secretary of Corps.
Ship left Sheerness for 3 hours commissioning trial.
At 2pm left for Portland in beautiful weather.
Maximum roll 2^0 and a following sea.
Very difficult to distinguish lights around the coast[11].
No lights other than steaming lights.

---

[11] Lighthouses, lightships and lightbuoys, whereas a sidelight was seafarers name for a porthole or window.

Friday 19th February

Good night's rest although for first half hour it seemed strange due to the throbbing of the engines and the wash of the sea against the sidelight.

Ships passing in the night.

Cliffs at Isle of Wight and St Albans Head very white in the sunlight.

Glorious morning on looking out through the sidelight to see the sun shining.

Steamed into Harbour at Portland at 11am.

Moored in the harbour near Lord Nelson, Triumph, Irresistable, Roxbrugh.
King Edward VII shown class 3 etc etc[12]

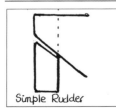 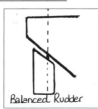 [12]. The eight ships in the King Edward VII class were called "The Wobbly Eight.". They had balanced rudders making them more manoeuvrable, but also more unstable, even straight ahead, hence the nick-name.

Saturday 20th February Spent most of morning looking at cliffs and prison from the ship and the general appearance of the harbour.

Ashore in afternoon to purchase maps and walk along promenade. Landing place at the Torpedo Jetty very difficult to land especially if there is any sea.

Haircut in a grubby little place. "All that glisters is not gold".

Sunday 21st February

Divisions in the morning was very monotonous waiting about on the Quarter deck. Ships company all inspected individually reporting to the Captain who found it difficult to impress on the men that they were reporting to him and not to the officer calling out the names.

After Divisions, fall out and at 10.30 most went to Church which I did not attend. I went ashore. Walked to Wyke looking for a Chapel but finally returned to Weymouth Wesleyan after inspecting the ruins of Sandsfoot Castle between Weymouth and Portland.

Returned on board 8.45 with Staff Surgeon O'Malley.

Monday 22nd February

Preparing lectures for Midshipmen. Ashore for a walk.

Tuesday 23rd February

Attempted a photo of breakwater when Brittania was steaming through.

Wednesday 24th February

Examined all the Watertight Doors in the ship for efficiency. All in good condition.

Thursday 25th February

Investment of £50 in Local Loan Stock, transfer from Post office deposit. Went ashore to meet *ex-colleagues* Dippy and W.G.Sanders at landing stage *from other ships*. Afternoon tea at Trocadero Hotel Weymouth. On board 6.45 to dinner.

Friday 26th February
Preparing lectures for Midshipmen.

Saturday 27th February
Ashore in afternoon.
The country being delightfully hilly and roads good for cycling.
Coast guard station at Fleet and to Linton Hills. Arrived at Abbotsbury 5.40 and found the village the neatest and cleanest that I have seen up to the present. Abbotsbury is noted for its 1200 swans where I met Brown and Gardner. The view from Linton Hill was magnificent just as the sun was setting over the waters and the valleys looking peaceful all around. Went on board 8.45 absolutely tired out.

Sunday 28th February
Woke up slightly tired and glad to rest. Did not turn out for Divisions nor Church.
Wrote up some notes on "The Tobacco Smoking Question" from Frank Ballard.
Went ashore at 6.30pm to Weymouth Wesleyan Church service and prayer meeting which lacked the Methodist spirit. Boat shoved off 6 minutes early so I had to wait 2 hours walking the Sunday night promenade, during which time I am afraid most of the benefit of the service was lost. Glad to get aboard after 2 hours walking on the promenade. Place quite desolate and dark on a Sunday night.

Monday 1st March
Received monthly pay.
Buying shares in Cunard Company.
Very lazy and tired.
Spent day reading up Seamanship etc.

Tuesday 2nd March
Snow and hail. Another very slack day with practically nothing to see on board and no inclination to go ashore.
Wrote to Chrissie asking if convenient to weekend.

Wednesday 3rd March
Still waiting for fine weather to get ashore for some exercise.
Ward room for most of the day.
Wrote home.

Thursday 4th March

Weather rather bad.
Lecture to midshipmen on Stability.
Examined Watertight doors and Draft Indicator.
Looking forward to the weekend.
Anxiously awaiting for reply from Chrissie.

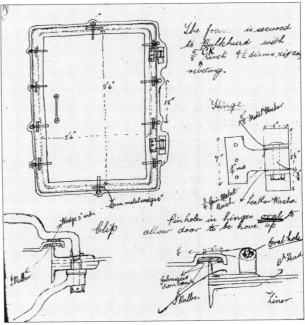

Friday 5th March

Waiting about looking for work.
Had a telegram from *brother* William *a shipwright, working and living in Plymouth* asking me for the weekend.
Left Weymouth 2.50 train via Yeovil, Taunton and Exeter to Plymouth at 8.30. Walked to Hotham Place was very glad to arrive and found Will was not ashore much to my annoyance.

Saturday 6th March

Did not see the sun rise but had a very full breakfast at 10am and then for a stroll on Plymouth Hoe and a walk through Plymouth Market.
After lunch went to Rugby match; Devon Albion 12pts, Leicester nil.
Returned to find Will had arrived.
Went to North Road to meet Amy but she did not arrive. In evening walked into Plymouth with Amy to order Devon cream to be sent home to Pembroke Dock.
Burnt the midnight oil. Retired 1 am Sunday.

Sunday 7th March
Rather late getting up but rather tired. Walk on the Lines with Amy in the morning and did not go to church. Got caught in a shower of rain for punishment.
In afternoon walk round Milehouse.
Saw Amy off 8.30 train and walked round Plymouth with Will and saw Monday morning in before retiring.

Monday 8th March
Very sorry weekend has come to a close.
Long journey back to Weymouth catching 11.10pm boat for Implacable, turning in at midnight very tired.

Tuesday 9th March
Very busy day.
Repairs to Watertight door in Engine room M L Bulkhead which was very stiff.
All double bottoms closed preparatory to sea going on March 10th.
Designing pendulum for measuring rolling.
Lecture in evening on Pumping and Draining to Midshipmen.
Preparations made for towing and being taken in tow.
Wrote to Mother.
General quarters[13] in afternoon.

---

[13] The Order "General Quarters" meant preparing for action and everyone had to go to the station he would occupy should an enemy be within sight. Seamen and marines manned the guns. Any Stokers not below in the stokeholds and engine room had to form fire-parties or stretcher-parties. Men not required at the guns went to shell rooms and magazines. The doctors and their staff prepared the operating theatre. Artisans went to shell hoists and magazine passages. The guns were readied and loaded with the exact weight of the real cordite or dummy practice charges. As each unit was ready for action, the officer in charge reported to the bridge.

Wednesday 10th March

Lecture on Stability to Midshipmen.
Hoisting in boom boats[14] by hand.

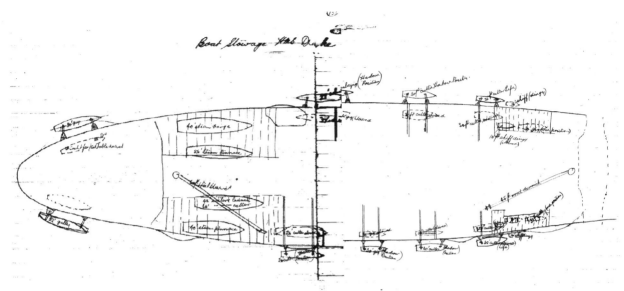

Weighed anchor at 1pm and left in company with Fleet for Lough Swilly.
Order "Collision quarters"[15]
I finished notes on cigarette smoking and was finally convinced it is not good and absolutely unscientific and unwise.
Letter from *sister* Edie.
Wrote to London & County Banking Group for Prospectus.
Fairly heavy sea running and sidelights quite awash but not seasick although waiting for symptoms to appear. I find the safest thing is to sit down.

---

[14] Launches and pinnaces were stowed amidships (along the centre), with the spare spars. A derrick was required to hoist the boats in and out. The smaller boats would be carried at davits at instant readiness whenever a ship was at sea for life-saving and general purposes. Albert's student sketch shows boat storage on HMS Drake.

[15] On hearing 'Collision Quarters' all crew had to go to an assigned post for life saving or 'Abandon Ship' routine.

Thursday 11th March

Awoke to find ship passing Cornish coast with very heavy sea.
Remained in bed most of the day with delightful headache which accompanies seasickness. Attempted to get up but finally decided to remain in bed.

Friday 12th March

Sea much better, steaming generally at 14 knots.
Tactics[16] in Irish Sea off Isle of Man.
Circling.

---

[16] "Tactics" were dictated by Fleet commanders, each Division and each ship obeying orders as the Admirals tried out manoeuvres and tested the efficiency of their Captains.

Chapter 3   Ireland and Scotland

Saturday 13th March 1909
Arrived at entrance to Lough Swilly at 10am and steamed to moorings off Fahon. Ashore at Fahon in afternoon 4pm and rode 8 miles *sitting sideways* in Irish Jaunting Car to Londonderry.
One Mile walk round walls and went into the new Roman Catholic Cathedral and round Brook Park.
Had Tea and Dinner in Northern Counties Club where Officers of the Fleet are Honorary Members.
Returned in Jaunting car to station and 9.30 train for Fahon.
Return to Implacable in cutter from Fahan pier and got very wet and cold.

Sunday 14th March
Snow and rain.
Remained on board all day.
Letter from Mother.
Reading and preparing Seamanship.

Monday 15th March
All watertight doors etc closed up for sea.
Reading and preparing Seamanship.
Unmoored 5pm and prepared for sea.
Starboard Cable Holder jamming.
Left Lough Swilly at 5.30pm.
Not seasick this time.
Letter from Mother.

Tuesday 16th March
Cruising in Irish Sea off Isle of Man.
Life cutter lowered and sent off.
Mails from Holyhead with scout.
Letter from Mother.

Preparing IHP [17] and revolution curves from Indicator diagrams.

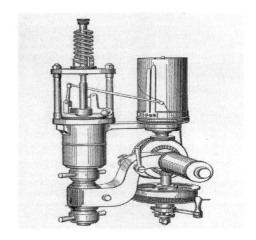
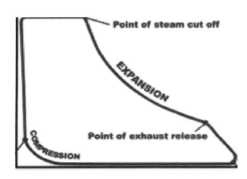

---

[17] The Indicated Horse-Power (IHP) of each engine was obtained indirectly from an Indicator Diagram, as compared with the actual Horse Power produced. The sketch shows an Engineer obtaining an Indicator Diagram for a small horizontal engine. The pen marks a graph on a paper chart wrapped around the drum. The drum revolves by a linkage that replicates the movement of the piston and pipework connects to the steam pressure in the engine. The graph records pressure versus volume. The area inside the graph is proportional to the Horse Power. The area of the graph is measured using a planimeter and knowing the revs, the Horsepower can be calculated. It needed a meticulous approach but was tedious and repetitive over a range of speeds and steam pressures for every cylinder on both engines.

Wednesday 17th March

Evolutions[18] directed from Flagship. "Forming line ahead and abreast".

Evolutions "Overboard Night Lifebuoys"[19].
One lost, second one broken. Cutter sent out to pick it up.
Printed off photos.

---

[18] "Evolutions" were competitive exercises ordered by the Admiral to his Fleet. After it was reported to him that every ship was ready, an order was given by wireless and flags. After the last ship hoisted the 'Answering Pennant' showing they understood the signal, the order was given "Haul down," and the men rushed away to complete their allotted task. Completion of the Evolution was signified by hoisting the finishing pennant on each ship as she finished.

[19] A naval life buoy was made with two connected large copper balls capable of keeping four men afloat. It was attached to a ship's stern by means of a slip, which was disconnected by pulling a trigger, so the buoy was immediately freed to fall into the sea. At night, the same trigger fired a friction tube, which ignited a fuse that exhibited a blue light, and burnt for twenty minutes, thus marking the position of the buoy. A sentry was constantly stationed by this buoy at sea. The buoy was primed every night at sunset, the tube being removed again at sunrise.

Thursday 18th March
Rather rough off Cornwall.
Prepared a report on smoke issuing from funnel.[20]

Arrived off Plymouth 6.30pm.

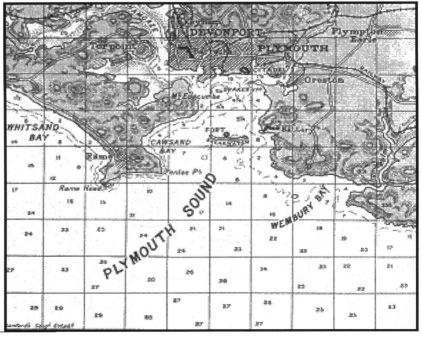

---
[20] Implacable's open bridge and the forward mast fire control platform, just ahead the funnel, could get smoke and fumes in a following wind.

Friday 19th March
Order "Test flooding of magazines". [21]

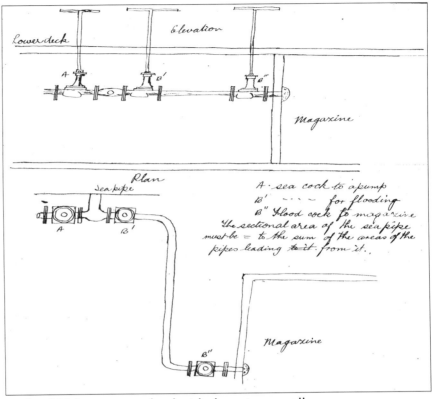

Examined Bakery. Capacity 900 loaves per day but bakery too small.
Arrived off Portland and moored in harbour at 6pm. Very glad to get back to harbour.

---
[21] Emergency flooding with sea water is forced in by pumps as in Albert's sketch.

Put ventilation of cabin flat in working order.
Letter from *brother* Harry *in Merchant Navy as ship's carpenter.*
Wrote to Institution of Naval Architects for 6 technical papers.
Orders that Implacable proceed to Firth of Forth on 24th March.

Saturday 20th March
Plotting curves of engines revs and IHP for Implacable.
Ashore in evening to purchase photographic gear.

Sunday 21st March
Letters from Mother and Harry.
George in Gibraltar and Harry leaving Amsterdam for London.

Monday 22nd March
Coaled ship 9.30am to 7.30pm, 950 tons.
Shrouds of collier carried away.

Tuesday 23rd March
Football match at Naval Recreation Ground. Final for Channel Fleet Cup.
Implacable 6 goals, HMS Dominion 2.
Walk round beach to Weymouth and returned to ship at 6.45.

Thursday 25th March
Ship left Portland at 8.30am arriving at Portsmouth No.7 buoy at 2pm.
Went ashore 4pm and purchased Tennyson's poems.
Landed at dockyard and had a walk round docks etc
Received shares of Cunard Co from London & County Banking C°.

Friday 26th March
Left Portsmouth at 10am.
Commenced 8 hours trial at full power. Average 18.2 knots, 108 revs, Calculation of power during trial 15000 IHP. Vibration at full speed very great. Also 16 hours trial at 3/5 power.
General quarters at 10pm.

Saturday 27th March

Rather seedy but had IHP calculations to complete.

Passed HMS Irresistable just outside Firth of Forth.

Arrived at Queensferry below Forth Bridge at 8pm and very glad.

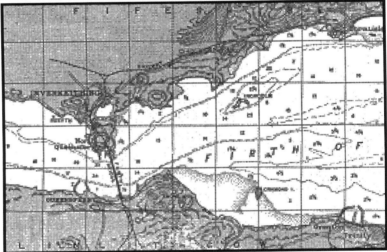

Interview with Captain who desired to see my professional journal of which none has yet been written up.

Order for General Quarters 'Fire Control' in Transmitting Station[22]

Sunday 28th March
Letter from Mother.
Very wet and dirty all day.
Read Tennyson's 'In Memoriam'.

Monday 29th March
Preparing quarterly account of Double bottoms.
Interview with Captain with respect to Journal.

Tuesday 30th March
Letter from Harry (London).
Edinburgh in afternoon by 2.5 train and spent afternoon in Royal Scottish Academy after walking along Princes Street Gardens, Scott's Memorial.
Tea at Littlejohns in Princes Street.
Returned to Dalmeny by 5 pm and walked round Queensferry.
Evening Captain dined in Ward room and entertainment on upper deck with Ship's Company.

---

[22] The Admiralty wanted to test the gun firing by Control from a centralised firing point on the mast, where range finders, gyroscopically stabilised sights, predicted target positions by mathematical plotting could be carried out and instructions relayed to the gun crews. A more sophisticated trial was being developed on other ships in the Fleet, such as HMS Bellerophon, called Director Fire which controlled all the guns centrally.

Chapter 4   A New Captain

Wednesday 31st March 1909
Captain Tottenham took over ship.
Fire Control to test gyroscopes.

Thursday 1st April
Photograph of Forth Bridge.
Letter from Edie with IBRA papers.
Wrote to Harry RA Docks and sent for developing card.
Edinburgh 2pm.
Walk to monuments Nelson, Wellington and City Observatory, overlooking the jail, past Burns' Memorial down to Holyrood Palace. Historical Apartments of Mary Queen of Scots. Large tapestries in most of the bedrooms. Walk round Queens Park overlooking the town, back to Princes Street via Parliament Square, St Giles Church, Academy, Royal College of Surgeons.
Returned to Queensferry 9.5.

Friday 2nd April
General Quarters.
Stationed in Fire Control Forward.
5 Officers 11 men.
Preparations for repair of Ash Empeller

Saturday 3rd April

Sketches of steering gear clutches for Engineer Commander.

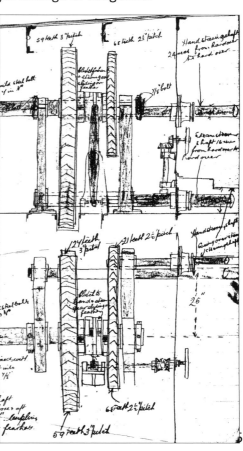

Diver sent down to plug hole for Ash Ejector[23], *but* tide too strong.
Ashore in afternoon, met Bentley at Hams pier and walked to Forth bridge.
Returned to ship 7.45 boat.

Sunday 4th April
Palm Sunday. Divisions 9.30.
Church 10.30 in Fore Mess Deck. Text "No man liveth unto himself". Roman 14-7. Either consciously or unconsciously every one exercises an influence for good or evil helping or dragging down those around us.
Letter from Mother.
Letter from London & County Bank.

Monday 5th April
Examined watertight doors.
Ashore in afternoon and cycled to Linlithgow via Newton. Snapshot of Linlithgow bridge. 8 miles. Returned on board 6.45 pm.

Tuesday 6th April
Examined all Watertight doors and ventilation arrangements, valves and fans.

Wednesday 7th April
Unmoored 9.30 am. Got under way 1.30 pm leaving Firth of Forth for Cromarty.
Exercises in the North Sea with Dreadnought, Lord Nelson, Invincible, Irresistible, Bulwark, Minotaur, Dido, Iris, Agamemnon, Implacable.

Thursday 8th April
Exercise in North Sea under Admiral May.
Met 2nd Division and Cruiser Squadron off Firth of Forth.

Friday 9th April Good Friday.
Sunday routine but did not have Divisions.

---

[23] Because the boilers were below the waterline, the waste ash and clinker had to be manhandled aft along the stokehold floor via a grill to an Ejector or Empellor to be forced out by steam power from the boilers as in cutaway diagram.

Saturday 10th April

Arrived at Cromarty Firth 8am. Letter from Harry (Glasgow). Ashore in afternoon 1pm. Met Dippy for walk to Fortrose (halfway) & returned to Cromarty to tea. Met Noble walking round Cromarty. Millais monument & Institute built by Andrew Carnegie in memory of Hugh Miller, geologist. Went aboard 7pm.

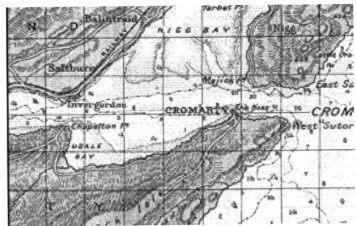

Sunday 11th April

Ships in Cromarty
Destroyers
2nd Division & Cruisers.
Minotaur flagship of 1st Cruiser Squadron.
Railway station (nearest) Invergordon

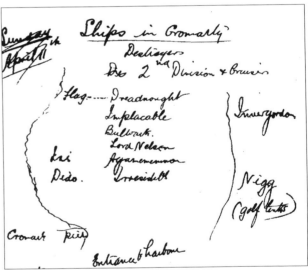

Noted Boats to ashore 1.00 3.45 6.30. Ferry from Nigg to Cromarty (golf links)

Monday 12th April
Trial of gyroscope[24] in Fire Control by sighting steam pinnace off ship.
Senior Wardroom officers dining with Captain.

Tuesday 13th April
Atlantic Fleet arrived at Cromarty 8am, also Bellerophon[25].

Ashore 1pm to meet Noble. Met Noble and had tea at Cromarty with Robertson, Gardener & Eng^r Lt Cooke.

    Purchased golf bag and club.

Very curious effect due to the reflection of sun from the snow on the mountains, everything seemed to be blue over the harbour.

Wednesday 14th April Letter from home.
Thursday 15th April
Left Cromarty 6am in company with Atlantic Fleet for Exercise in North Sea under cloudless sky. Exercise in "Control Stations"

---

[24] Probably a test to cross check the range and position of a target (pinnace) at a measured distance and to transmit the required trianing and elevation of guns to the firing officer, who would have to allow for pitch, roll and yaw of the ship.

[25] HMS Bellerophon, with distinctive strengthened masts to support large fire control platforms, had just been commissioned. Significantly, her guns were hydraulically controlled so that they continuously aimed at the target, adjusting for pitch, roll and yaw.

Friday 16th April
Exercises in North Sea and then orders to Scapa Flow, Orkney Islands

Saturday 17th April
Arrived Scapa Flow. 10am. Moored about 2 miles from shore.
Home Fleet, Atlantic Fleet and Destroyer Squadron all in Scapa Bay.

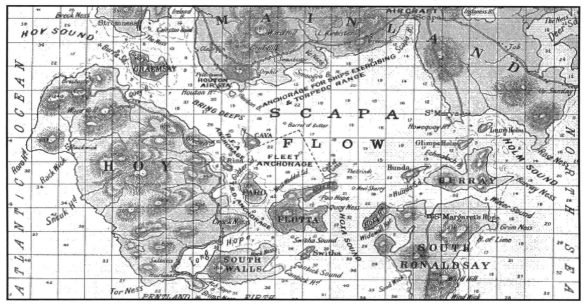

Ashore in afternoon and walked to Kirkwall. Met Noble and climbed to top of old Cathedral outside of Belfry. Walked round Coast Guard Station.

Sunday 18th April
Ashore 1.30. Met Noble and walked to St Mary's and got wet through.
Returned 7.30.

Monday 19th April
Letter from home.
Raining all day.
Calculations on flooding compartments.

Tuesday 20th April
Coaled ship 5.30am till 2pm. 620 tons.
Examined doors etc
Calculations for flooding compartments.
Wrote to Harry and sent for marine glasses. Letter from Jack.
Fleet as moored in Scapa Flow.

Wednesday 21st April
All ships except 1st Division left for Cromarty.

Thursday 22nd April
Unmoored 7am and steamed out off Hoy.
Laying out Number 3 range for target practice small guns.
Tow Target. Practice with small guns .303

Friday 23rd April
Tow target for .303 with 12 pounder and 3 pounder.

Saturday 24th April
Tow target for .303
Returned Scapa and moored 5.30pm.
Moving Target towed by 56ft launch past stationary moored ship.

Sunday 25th April
Church on quarter deck.
Remained on board all day.

Monday 26th April
Unmoored at 8am & proceeded to Range for 3 Pounder & 12 Pounder. Fixed targets but ship travelling knots.
Letter from home.

Tuesday 27th April
Left 5am for Pentland Firth to lay Triangular Targets and lights for night firing.

Wednesday 28th April
Night firing 12 Pounder and 3 P^r.
Letter from Will.

Thursday 29th April
Towed target and Night firing 12^{Pr} shell

Friday 30th April
Ship left Pentland Firth for Scapa Flow.

Saturday 1st May
Tested Light Quick Fire, 12pounder and 3pounder guns on No.3 Ranges.
Had to abandon firing due to swell.
Brother Harry left in S.S. Pakling, Blue Funnel Line

Sunday 2nd May
Ashore 1.30. Walked to other side of Island, returned 7.15.

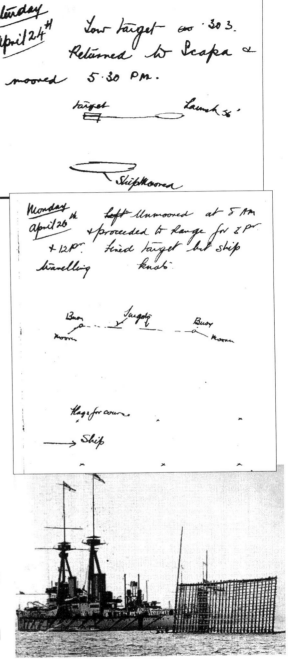

Monday 3rd May
Test firing 3 P^r and 12 P^r guns. Hits 91.

Tuesday 4th May
Practising moving targets.
Gun Mountings

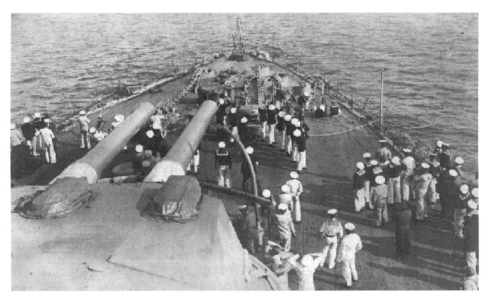

Wednesday 5th May
Brother Harry to join HMS Glasgow.
Brother Will sailed for Mediterranean in HMS Medea[26].

---

[26] A cruiser built for the tropics sheathed in wood and copper.

Thursday 6th May
Received cap covers from home. Received cap covers from Newcomb. Reading up seamanship.

Friday 7th May
Tow target practice in Scapa Sound

Saturday 8th May
Cycled to Stromness, leaving Kirkwall 1.45 arriving Stromness 3.45.
Tea at Mackays hotel.
Took photo of harbour and shipping.
Bought postcards to send home and to J.W. WBC and Dick Adams.

Left Stromness 4.45 Kirkwall 6.45pm

Sunday 9th May
Church on quarter deck. Text "Thou shalt love thy Neighbour as Thyself".
Wrote *to brother* George.

Monday 10th May
Letter from home.

Tuesday 11th May
Coaled ship 190 tons
5.30 -11am.
Ashore in afternoon with Robertson to Kirkwall met Noble on Scapa Pier.

Wednesday 12th May

Left Scapa 9.30am. arrived Cromarty about 5.30pm anchored off Cromarty Pier. Exercise in *Fire* Control on way South with Bellerophon steaming separately.

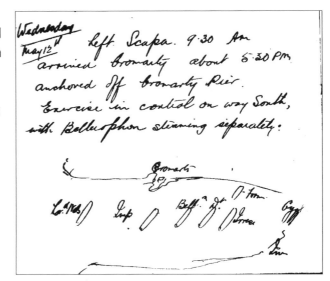

Thursday 13th May

Ship left Cromarty for Lossiemouth 7.30am for heavy gunlayers practice.
Sea very high and practice abandoned. Anchored off Lossiemouth.
Torpedo Boat Destroyer brought mail from Cromarty.

Letter from Hiraga[27]

Friday 14th May

Anchored off Lossiemouth. Sea still too high for practice.

Saturday 15th May

Returned to harbour in evening for stores etc.

Sunday 16th May

Left for Lossiemouth to lay down range for Heavy Gunlayers test.
Went back to Cromarty in evening. Letter from home.

---

[27] Hiraga and Albert were student friends at the Royal Naval College, Greenwich 1905-1908. When they graduated in 1908, Hiraga spent the next six months touring various shipyards in France and Italy before returning to Japan in early 1909. Later he became Head of Japanese Naval Construction. Albert built HMS Ajax, from which his nickname is derived. It is not known if any of their ships fought each other in World War II.

Monday 17th May
Left Cromarty for range *but* Heavy Gunlayers test abandoned on account of sea.

Tuesday 18th May
Finished Gunlayers' test and returned to harbour.

Wednesday 19th May
Ashore in afternoon at Cromarty. Walk around cliff.

Thursday 20th May
Left Cromarty with Admiral May on board for Battle practice in Moray Firth. Control exercise and firing from each heavy gun.
2 torpedoes fired from each tube with target in tow of steam pinnace. One torpedo turned in circle near to ship demonstrating the possibility of ship overrunning a torpedo fired from forward tube. Orders for firing "Stand by, Flood Tube, Out Bar, Fire".

Went into harbour at noon when Admiral May returned to Dreadnought.
Torpedo practice again carried out in afternoon.

Friday 21st May
Control exercise in Moray Firth.
Taking up running buoys from range.
Returned to harbour 4pm.

Saturday 22nd May
Ashore 10.30am with bicycle. Cycled to Inverness via Fortrose. Cromarty to Fortrose 10 miles. Fortrose to North Keswick 9 miles. Ferry crosses between North and South Keswick every hour. Lunched at Union Hotel. Main streets lined with hard pebbles, very rough surface. Several bridges crossing the river and high hills all round the town. Glorious day. Left Keswick ferry 4pm for Cromarty against head wind.
Arrived Cromarty 7pm very hot and tired.

Sunday 23rd May
Remained on board all day. Received a signal from C in C for accommodation for 3 more Assistant Constructors.

Monday 24th May
Left Cromarty with 1st and 2nd Division of Home Fleet at 9am.
Tactics in North Sea.
Had a surprise General Quarters at 10pm.

Tuesday 25th May
Commenced full power trials but had to be abandoned on account high pressure piston ring problems *on the triple expansion vertical engines.*
3/5th power at 15 knots carried out from 12 noon to 8am on 26th.
Very dirty, not quite seasick.
Working out Indicator diagrams.
Applied for 10 days leave.

Wednesday 26th May
Arrived off Sheerness 9.30am.
First Watch went on leave 5.30.

Thursday 27th May
Ashore and called Tom Adams at Sheerness Dockyard. Left 4.20pm, left Paddington 9.15

Friday 28th May
Arrived home *(60 High Street)* Pembroke Dock 7.30 am.

Chapter 5 The Naval Reviews

Monday 7th June
Left Pembroke Dock 7.35am. Arrived Sheerness 7.35pm. No boat till 11pm. Dinner at Fountain Hotel with Janiou. Back on board HMS Implacable. Letters Harry & Chrissie.

Tuesday 8th June
Implacable left Sheerness 1.30 pm with 1st Division of Home Fleet. Arrived off Dover at 6pm.

Wednesday 9th June
Joined Atlantic Fleet at 5.30am in English Channel under the Flag of Prince Louis of Battenburg.

Arrived at Portsmouth 3pm.

Thursday 10th June
Ashore 1.30 Victoria Pier.
Walked round Portsea and Southsea.

Friday 11th June
Preparations for Naval Review. Cleaning ship.

Saturday 12th June
Pleasure craft by the dozen cruising round the fleet.

Guests arrived on board by 12.30. Luncheon Party.
At 1pm Officers Call sounded. Admiralty Yacht 'Enchantress' had left Portsmouth and proceeded to steam through the lines.
'Enchantress' steamed first followed by 'Volcano' with Press representatives on board

and finally by Admiral Superintendent Portsmouth Dockyard and Yacht with distinguished Flag Officers. Rained at 2.30 and our guests left at 6 pm.

Sunday 13th June
Ashore in afternoon 1.30. Bible Class at Buckland Wesley Church Portsmouth. Walk round Copnor and Cosham and back to Pembroke Road Wesleyan Church Portsmouth.

Monday 14th June
General Exercise [28]. "Out Torpedo Nets"
"All boats crews out pulling"[29].

Left Spithead 4.30pm for Oban, Scotland.

Tuesday 15th June
Control Exercise in forenoon.
Testing night lifebuoys.
Tactical exercise

Wednesday 16th June
Control Exercise. Tactical.
Towing ship. HMS Cornwallis towing Implacable.

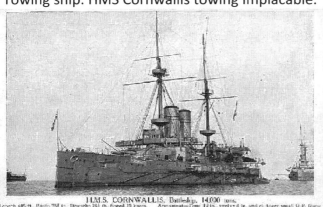

First lecture to new Midshipmen on Beams, Frames, Bending Strains.

---

[28] An order for all ships in Squadron or Fleet to conduct a non-competitive task.
[29] The Navy does not "row" a boat: it "pulls": so to pull is to row, and a pulling boat is a rowing boat.

Thursday 17th June
Arrived at Oban at 1.30 pm.
Mediterranean Fleet in harbour.

Friday 18th June
Small craft, Torpedo Boats, Destroyers, Scouts etc arrived at Oban.
Second lecture to midshipmen on Deck plating, Steam, Steering etc.

Saturday 19th June
Ashore at 1.15.
Cycled along coast to Melfort. Lot of small hills and round edge of two lakes. Returned at 7.30.

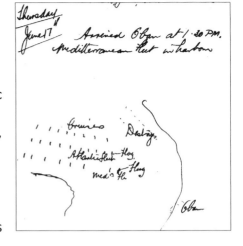

Sunday 20th June
Remained on board after Divisions.

Monday 21st June
Atlantic Fleet
Admirals' Inspection at 9.30
Officers mustered by Navy list on Quarter deck for Admirals inspection. Afterwards men were inspected. Vice-Admiral Prince Louis of Battenberg. Flagship HMS Exmouth Vice Admiral's Flagship Exmouth.

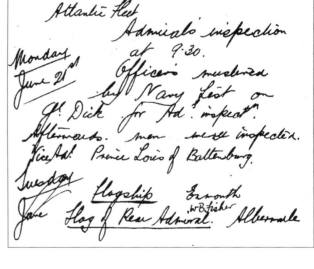

Flag of Rear Admiral W.B. Fisher HMS Albermarle.

Tuesday 22nd June
Very slack.
Received orders to be prepared for coaling night and day on account of impending manoeuvres.

Wednesday 23rd June
Coal ship 6am to 12 noon. Sweep collier.

MEN OF H.M.S. "IMPLACABLE" AFTER COALING OPERATIONS.

Thursday 24th June
Ashore 1.15 for golf on 9 hole course.
Foursome. (Fleet Surgeon 56 Self 68); (Paymaster 58 Lieut ?? D&E 59)

Friday 25th June.
Ashore 1.15.
Golf on 9 hole course. A.A. 1st round 64. 2nd round 59. Partner 39.

Saturday 26th June
Astn Constructor, E.B. Charig, appointed to HMS Implacable with me, to date 19th July

Sunday 27th June
Church on Quarter Deck. Remained on board all day.

Tuesday 29th June
Ashore 1.30. 3 rounds of Golf at Oban.

Wednesday 30th June
Lectures to Midshipmen 1.20 -3.30 on Steering.
Ashore 4 pm. Walk to Dunstaphon Castle.

Thursday 1st July
Ship cleared for action.
Ship darkened, no lights showing outboard. Preparation for night defence.
War Declared 11.35pm (Manoeuvres).
Cruisers left at midnight. Atlantic and Mediteranean fleets left at 12.30 midnight.
Night comparatively light. Dressed and went on deck at 1am but nothing to see except ships steaming in line.

Friday 2nd July
Fleet steaming for north coast of Ireland at $12^{1}/_{3}$ knots. Control exercise. Position off Eagle Island NW coast of Ireland *as in sketch*. Blue Fleet, Atlantic and Mediteranean steaming in line.
Battleships of Red Fleet sighted astern[30].
Red Fleet steaming up opened broadside fire on HMS Triumph and was put out of action and returned to Oban. Two of our protected Cruisers were returned to Cruiser Squadron.

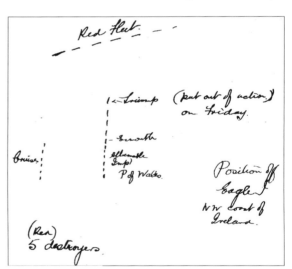

[30] HMS Prince of Wales was leading, HMS Triumph at rear of line.

Saturday 3rd July
Ship manned and armed till 5am.
About 5.30 HMS Bellerophon discovered on starboard side[31] after fog had lifted and put out of action.
Bellerophon had been engaging several cruisers of Blue fleet when enveloped in fog.
Joined up with White Fleet at 10pm thus making
8 Battleships White Fleet,
10 Battleships Blue Fleet
under Admiral Curzon Howe.

Sunday 4th July
7am General Quarters "Engagement with Red Fleet" resulting in victory for Red Fleet.
Control arrangement for plotting broke down. Speed indicator disconnected.
Everything over by 8am & fleets proceeded to their respective rendezvous.
When off Bantry Bay received message that Atlantic Fleet not to proceed to Oban but divert to Foynes in the Shannon.
Owing to difficulty of navigating *in fog* it was necessary to go into Dingle Bay at 7.30 and drop anchor.

Monday 5th July
Left Dingle Bay in the forenoon 10am and arrived at Foynes 8pm.

Tuesday 6th July
Ashore at 1.15 and left Foyne for Limerick at 2pm by train. Arriving Limerick 3.30 to see Cathedral and Carnegie Library. Tea in George St.
Returned by 6.25 to Foynes.

---

[31] Sketch confirms Prince of Wales leading the line astern which concentrated fire onto Bellerophon on its starboard side.

Wednesday 7th July
Instructing midshipmen 1.30 - 3.30.
Ashore 3.45 for walk round country to Foyne. Climbed hill and took photo of town and Cross erected to memory of the landlord by his tenants.

Thursday 8th July
Implacable left Foynes 5pm for Bantry Bay to calibrate.
Very dirty night.

Friday 9th July
On account of the fog, anchored off Berehaven early in the morning.
Arrived Glengariff 2pm.
Lectures to midshipmen 2.00 - 3.30 on Pumping and Ventilation.

Saturday 10th July
Ashore at Glengariff. Chief attraction two large hotels and soon returned on board.
Explored journey to Killarney 17 miles. Coach to Kenmare & then by train.
Ship left Glengariff 7.30pm for Queenstown[32].

Sunday 11th July
Arrived Queenstown 6am. Other ships in harbour, including 4 Cruisers.
Ashore at 4.30 walk round and Wesley Chapel in evening.
Another walk round country and returned on board 10.30 pm.

Monday 12th July
Coal ship 5.30am. (Very foolishly got up 6am & remained most of time on collier) Finished coaling 11am. 686 tons.
Ashore at 1.30 at Queenstown

Tuesday 13th July
Ashore 1.30 to play 2 rounds at Monkston 9 hole golf course.

---

[32] Later renamed Cobh or Cork,

Wednesday 14th July
Lecture to midshipmen on Coaling etc.
Unmoored and went outside harbour at 7.30pm to anchor.

Thursday 15th July
Joined up with the Atlantic Fleet 5am and proceeded to Southend.
Evolution "Tow Ship".
Implacable towed HMS Russell for 1 hour at 50 revs.

Friday 16th July
Steaming up Channel.
Instruction to Midshipmen (Sketching).
After evening quarters, Evolution "Out Collision Mats"[33].

---

[33] Canvas Collision Mats were made of very stout canvas fitted with ropes at each corner. The ropes enabled crew members to manoeuvre the mat into position under the ship where the hull was damaged. It was held in place by the force of the water but, only slowed the inflow and could not completely stop the water.

Saturday 17th July
Passed HMS Edinburgh[34] in Channel and arrived at Southend 3.15pm.

Sunday 18th July
Church at 10.30 Quarter Deck. Ashore at 1.30 at Southend and Southchurch.

Seafront & pier very crowded. Took 1 hour to get from one end of pier to the other on account of the crowd. Returned 7pm.

---

[34] This weird looking ship had been built in Pembroke Dock but in her last days was converted to a target ship with modern armour plates. The Navy wanted to measure the effect of armour-piercing shells filled with Lyddite, the most potent explosive of the period, on these plates.

Monday 19th July
Colleague Charig arrived on board at noon.
Ashore 1.15 at Southend. Tea in Winter Gardens at the Palace Hotel. After buying Postcards etc, walked along Promenade to Westcliff.

Tuesday 20th July
Ashore 2pm at Southend.
Tea and walked to Leigh-on-sea.
Had to wait on pier till the Mayor's Party had gone to pier before returning on board 7pm.

Wednesday 21st July
1200 men of the Fleet went to London to lunch and receiving present of pips etc etc.
Lecture to midshipmen on Anchors and Cables.

Thursday 22nd July
Ashore at Southend 3.45 met Noble, had tea together at Hotel Metropole.
Tram to Leigh-on-Sea and walked back along the Promenade. Supper and then went on board 10pm.
Fleet illuminated. Each ship outlined by electric lamps along the waterline, rails, masts, funnels and bridge.
Took photo from bridge.

Friday 23rd July
Court of Inquiry on board HMS Implacable.
President of Court Rear Admiral W.B.Fisher.
Concerning condition of hydraulic machinery and gun mountings.
Enquiry due to Gunnery Lieutenant condemning the fittings as fitted by dockyard during quadrennial refit 1908-09.
Dockyard represented by Engineer Rear Admiral Rudd.

Saturday 24th July

Left Southend 6.30am for Dover. Passed wreck on Goodwins to port on voyage down Dover.

Took measurements of wave times and lengths[35].

---

[35] Albert was aware that one of the observations Admiral Beresford required in his Order was the measurement of wave time and length. All Naval Constructors knew from elementary theory that there was a critical wavelength of a ship to give a maximum back breaking stress. This reversed a half wave later, from hogging to sagging. The calculations assumed minimum midship loading, typically fuel, for hogging and maximum loading for sagging as in Albert's lower sketch. In May 1910, soon after leaving Implacable he made these sketches and notes during trials of HMS Lion.

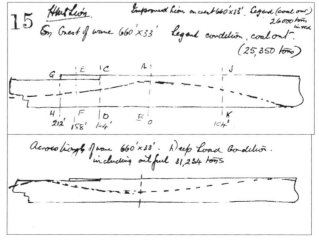

The first visit of the Fleet to Dover in March 1909. HMS Prince of Wales moored at the Prince Of Wales Pier.

Arrived at Dover 11.45 in very strong wind rendering mooring difficult, so proceeded to buoy but missed at first attempt and had to go outside and then back again. Ship was not moored till 2pm.

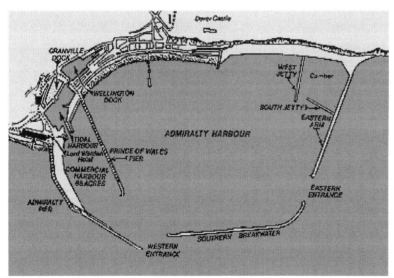

Ashore 2.30 Prince of Wales Pier and walk round Castle. Went to Church in the Castle. Saw Queen Elizabeth's pocket pistol 24ft long, brass piece of ordnance about 5" bore. Walked through Connaught Park, had tea in town.
Walk to station and returned back on board 7pm.

Sunday 25th July

Church on quarter deck. Monsieur Bleriot crossed the channel from Calais to Dover about 5am taking just over 20 minutes. The monoplane, the first to cross the Channel alighted at rear of Dover Castle on the golf links and broke the front wheels on striking the ground.

Ashore in evening 5.45 and after service at Wesley Chapel walked to castle & saw aeroplane.

Supper at the French restaurant in Street.

Monday 26th July

Left Dover 9am and had control exercise in Channel. Plotting for cross channel steamers. Returned harbour 3pm. Ashore at Dover 3.45 Walk round Shakespeare Cliff.

Tuesday 27th July

Left Dover 9am and carried out Turning Circle Trials in Channel, taking times for every point turned for plotting purposes. With Charig, measured heel due to circle on timing and also the time to get the helm over at various speeds 10, 12, 14 knots. Returned to Dover at 3pm.

News received at 5pm that M Latham was to make second attempt to cross the Channel on his aeroplane. Excitement was intense at 5.52 when news was received that he had started. French destroyers were disposed in various positions across the Channel. The aeroplane was sighted at 6.10 appearing as a speck and very rapidly approached Dover looking like a huge dragonfly. About a mile from the shore-line the aeroplane swooped down rather rapidly and then appeared to recover itself but finally plunged into the sea. Ships' picket boats and French Destroyers steamed to the spot where the aeronaut could be seen standing in the machine. From the ship the propeller could be seen revolving and it was due to the motor breaking down that the failure occurred. Mr Latham was taken aboard a French Destroyer to Dover harbour.

Wednesday 28th July

Left Dover 9am for Control exercise in Channel. Very good results by plotting the cross channel steamers about 20 knots. Returned to harbour 3pm. As result of Captains order, I took 1 dog watch[36] in engine room, grossly unjust and unfair.

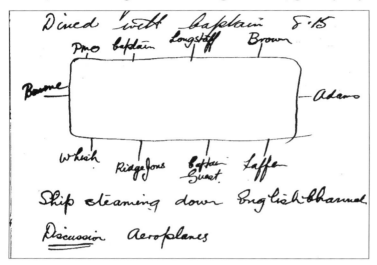

Left Dover 7pm in company with First Division of Home Fleet and Cruiser Squadrons. Dined with Captain 8.15. (sketch)

Ship steaming down English Channel.

Discussion "Aeroplanes".

Thursday 29th July

Passed German Fleet of 16 Battleships and 10 Cruisers in the English Channel. Some conspicuous by large derricks making a good target.

Arrived at Cowes 1pm and anchored in 5 lines in preparation for the Review.

---

[36] The day was divided into 5 four hour watches and 2 two hour dog watches. If the crew was split into two halves and worked alternate watches the effect of including one dog watch in 24hours meant a sailor was not always on duty at the same time each day.

Friday 30th July
Lecture to Midshipmen on Stability.
Ashore at Cowes afternoon.

Saturday 31st July
Review by H.M. the King in yacht Alexandra. 2.45pm.
No guests allowed on Quarter Deck during Review.
Officers in Full Dress.
Ships illuminated at 9pm.
Officer of the day.

Sunday 1st August 1909

Signal to wear white trousers and boots for the first time this year and the first time of my wearing them at all. Church on board.

Ashore 2pm crossed by ferry to East Cowes and walked to Osborne and Whippingham Church. Returned and had tea at Prince of Wales Hotel just outside grounds of Osborne House. Returned to West Cowes and attended service at Wesley Chapel 6.30. After Church walked to Gurnard 2 miles from Cowes.

Monday 2nd August Bank holiday.

Reviewed at 3.45 by HM the King and the Tzar of Russia.

British Torpedo Boat Destroyers cleared the lines followed by Kings Yacht, Alexandra, with King and Tzar on the bridge.

Next 2 large Russian Cruisers steamed through followed by 2 Russian Destroyers

Point of interest on Russian Cruisers. Wireless Telegraphy Arrangement on the Foremast Main Mast.

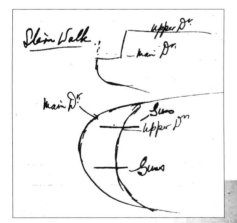

Another point of interest was the Stern Walk design for guns on upper deck.

Owing to the length of the Russian Cruiser and unhandiness it was impossible for them to turn in between the lines of the 1st and 2nd Divisions of the Home Fleet.

Cruiser (Rurik) had to manoeuvre for about 5 or 10 minutes before getting through.

Inserted is Alexander Spiridovitch's account from the Russian perspective. One could see in the distance along the coastline a forest of masts. That was Portsmouth. On the left, we could see the coast of the Isle of Wight, The strait which separates Portsmouth from the Isle of Wight forms the harbour of Spithead, unique in the world. We had before us the entire North squadron of the English fleet. Three lines of huge combat ships and many lines of smaller ships were arranged in parallel in the harbour of Spithead and were lost out toward the direction of Cowes. One hundred Fifty-three of them, without counting the destroyers and the smaller ships, commanded by 28 admirals, who were receiving their crowned admiral, the Emperor of Russia.

A characteristic "hurrah", which was similar to our Russian "hurrah", came to us from the ships also along with the sounds of the Russian national anthem. Our sailors responded back at full voice. We passed before ships each more and more impressive. We arrived before the right flank where we found many Dreadnaughts, the pride of the British fleet. This type of ship was at the time a novelty to us, as we did not yet have even one of this class of ship. Seeming like gigantic and monstrous irons as we passed by, they pressed down, so to speak, compressing the entire surface of the sea.

The yachts returned along the same route in reverse and passed before the second line. On several of the ships we could see women. This astonished us. The women were waving handkerchiefs. It was not without some feelings of jealousy that we watched this admirable tableau. If only we had something similar!

Passing before the Dreadnaughts, the Rurik could not turn as required, could not "deploy" and failed to hook up with one of the ships. It had to execute a manoeuvre which had the effect to make the ship leave the line. It soon joined back up with the rest of the squadron, at the same place it had occupied before the mishap.

The parade of ships lasted more than an hour. Then at 5 o'clock, the yachts returned to their places and dropped anchor, and one of the Dreadnaughts began to salute them with cannon shots. The monster made an indescribable thunder. Before us and to our left the entire surface of the sea was covered with yachts and small boats of all kinds. A genuine forest of masts it was, with flags flying from their tops.

Farther on, on the coast, we could see Cowes, where they were then holding the annual Royal Yacht Club races. All of the sporting members of English high-society were gathered there. The attention of yachtsmen around the world was concentrated on Cowes.

Our passengers were pressing, enervated, each wanting to disembark for Cowes as soon as possible. They asked impatiently why they were late in going into the boats.

The handsome sailor of the Equipage of the Guard who was the pilot of our motorboat, asked Prince Byelozerski permission to cast off. Byelozerski made an approving nod of his head and we left.

The boat went well, but then the Prince gave the order to the pilot to go in a straight line to the quay which would be faster. The pilot respectfully observed that he had to follow the route. The Prince became angry and insisted and gave him a military order. The sailor obeyed. We went out for a few seconds in a straight line, and then we heard a dreadful crash and the boat stopped. The pilot had been right. The Prince was really mad. His companions told him to stop getting involved in things he did not understand, and that the pilot knew full well what to do and how it should be done.

We back up, we went forward, and did so several times. We finally succeeded in freeing ourselves and went into open water, and after several minutes we neared the shore without further incidents, and finally to the quay. Cowes was crammed with people. They looked at us with curiosity. They admired the handsome sailors of the Equipage of the Guard who were truly magnificent.

When we awoke the next morning, the Fleet was no longer there. Silently, without anyone having noticed, they left the harbour during the night. Only a true sailor can really appreciate the virtuosity of such a manoeuvre.

Chapter 6   Battle Practice

Tuesday 3rd August
Left Cowes at 5.30 am for Berehaven to calibrate and work up for Battle Practice.
Control exercise.
Took afternoon watch.
Evolution exercise "Away boats crews"[37].

Wednesday 4th Aug
Control exercise. Afternoon Watch.
Hockey on quarter deck in evening.
Evolution "Let go night life buoys".
Arrived Bantry Bay and moored off Glengariff 8pm.
HMS Hannibal at calibrating range had to give place to Implacable on account of working up for battle practice.

Thursday 5th August
Shifted billet to calibrating range at 5am near Medusa.
Calibrated 12" and 6" guns, 3 rounds per gun. Muzzle velocity is measured by firing through the 2 frame screens at a certain known distance apart and the time shot takes to pass

HMS Medusa

between the frames gives the velocity. The Medusa has two masts rigged near ends of the ship. The frames are suspended from yard arms, the frames being in circuit with a chronograph.
The Calibrating Officer of Medusa was on board during firing.

---

[37] When a boat is "called away", the crew man their boat, put on cork jackets, sit on the thwarts until the Coxswain, following a strict set of rules, orders the lowering of boat from davits and pull away from ship.

After calibration, steamed 10 miles down to Berehaven with Starboard engine only as port engine being under repair. At 72 revs it was necessary to keep 12^0 of helm. In considering resistance with only one engine working it is necessary to take account of extra resistance of propellor being dragged through the water.
Moored in Berehaven 8pm.

Friday 6th August
Picnic to Dunboy Castle grounds about 3 miles from ship, in cutter towed down by Picquet boat.
Bathing on beach before tea. Hornets nest near to picnic spot.
Walked round to Castle where a new wing had commenced but *was* left unfinished for lack of funds.

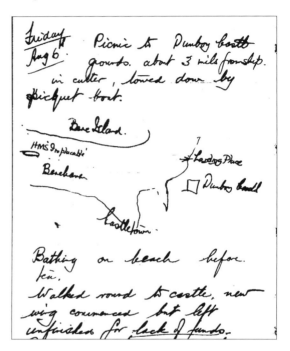

Saturday 7th August
Preparation for Battle Practice Control.
Left harbour 8am and returned 5.30pm.

Sunday 8th August
Church on quarter deck in morning. Ashore in afternoon in cutter to grounds of Dunboy Castle for tea on the beach. Annoyed by 2 hornets. Note. In most cases sting of 2 hornets sufficient to kill a man.
Had to row back to ship owing to Picquet boat running aground.

R N Steam Picket Boat

Monday 9th August

Control and battle practice with HMS Formidable Firing in the morning with Formidable towing and spotting.

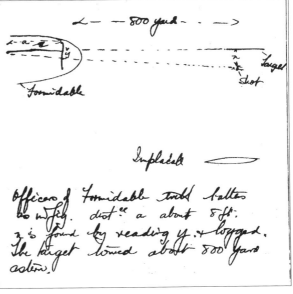

Officers of Formidable with battens as in figure. Observer at known, distance 'a', about 8ft *from sighting board* reads 'y' *as shot hits water*. The target towed about 800yards astern. The shortfall distance 'x' is found by reading 'y' and logged.

Tuesday 10th Aug

Control and Battle Practice exercise outside harbour with HMS Formidable. Other ships working up for battle practice in harbour were Queen, Prince of Wales, Arrogant, Drake.

Ashore at 6pm with Carter, Asst Paymaster, landed at Furious pier and walked to Castletown. Dinner at Hotel. Returned on board 10pm.

Wednesday 11th Aug

Control and preparation for Battle Practice with HMS Formidable outside harbour. Returned to harbour 5pm.

Left Berehaven 12.30 midnight for Queenstown to land a patient for hospital treatment.

Thursday 12th August

Arrived Queenstown at 9am and landed patient at Haulbowline naval hospital.

Coaled ship in afternoon from two Temperley[38] lighters. Each lighter with 2 Transporters. Commenced coaling at 2pm and finished at 7.30pm having loaded 465 tons.

Left Queenstown 8pm for Berehaven.

R.N. Hospital Haulbowline

---

[38] The Temperley Shipping line was a private company experimenting with mechanised loading shown here ship to ship.

Friday 13th August
Arrived off Bantry Bay 4am in thick fog so waited for Battle Practice exercise, but fog too dense and returned to harbour at 2pm.

Saturday 14th August
Rowed ashore in skiff and bathed from shore with Lts. Whish, Brown and Adams.

Sunday 15th August
Remained on board all day. Snapshot in afternoon.

Monday 16th August
Left Harbour 8am for Battle Practice. Very monotonous plotting but results fairly good.
Working with HMS Formidable as regards towing the target.

Tuesday 17th August
Left Berehaven 8am. Plotting again for Battle Practice and returned to harbour at 2pm so went ashore with P. Had 3 rounds of golf on 9 hole course. Very poor links.

Wednesday 18th Aug
Preparing for Battle practice. Tested Sights near lighthouse in harbour, near to sunken collier.
First lecture to the new class of midshipmen on Stability.

Thursday 19th August
Ashore at Bere Island for 2 rounds of foursome golf with Chief, Lt Croker, Charig & Adams. Won by latter. Tea at Cottage.

Friday 20th August
HMS Prince of Wales firing at Battle Practice. 15 hits. HMS Queen 6 hits
Conditions: 4 minutes allowed for firing on each run. Returned to harbour 6pm.
Lecture to midshipmen on Structures, Stress, strain etc.

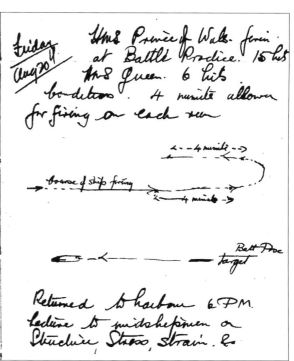

Saturday 21st August
Left harbour 8am.
Battle Practice Firing exercise at 4.30pm as below. Officers inspecting Battle were Rear Admiral Fisher, Rear Admiral Pearce ITP, Captain etc.
Target at 7 knots towed by Arrogant.
4 minutes allowed for firing on each run, one run on Starboard side and one on Port as in Fig.
6 hits 1 hit with 12" and 5 hits with 6".
Ship was yawing about 5 degrees on each side and rolling 3 or 4 degrees during the firing. Water being shipped on Fx and Quarter decks.
Returned to harbour 8pm.

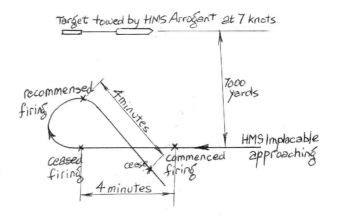

Sunday 22nd August
Church on Qtr deck 10.30am.
Ashore at Furious pier. Walked halfway to Adricole and then rode in jaunting car. Tea at Adricole Restaurant. Village dancing outside of the grounds. Walked back to Furious pier for 8pm boat.

Monday 23rd August
Ashore 1.30 at Bere Island Recreation Ground for Cricket match. Ward Room vs Gun Room, which Gun Room won by 1 run. I took 1 wicket for 7 runs. Tennis and Rounders etc.

Tuesday 24th August
Left Berehaven 6am for Sheerness carrying out 8 hours full power and 10 hour 3/5th power trials. 17.6 knots for 106 revs at 15000 IHP. Satisfactory.

Wednesday 25th August
Finished steam trial at 7am and proceeded to Sheerness at 13knots. Moored at No. 10 buoy 6.15pm.
Officers 1st watch went on leave.

Thursday 26th August

1st watch men went on leave at 9am. Ashore at Sheerness 1.30pm. Walk to Minster along cliffs. Returned on board 8pm.

Friday 27th August

Tennis at Home Fleet recreation ground. Robertson, Robinson, Longstaff. Returned on board 1.30pm.

Saturday 28th August
Left Queenboro Pier 6.25am boat train for fortnight's leave to home in Pembroke Dock. Arrived 6.10pm.
Harry also on home leave with *wife* Jennie.

Pembroke Dock

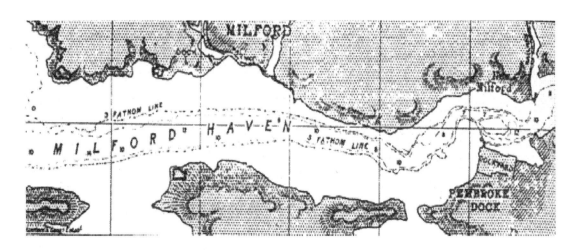

Chapter 7   Admirals' Inspections

Friday 17th Sept 1909
Joined ship in Sheerness overnight.
Hockey match. HMS Agamemnon 4 goals, Implacable 3 goals.
Met Tom Adams in Sheerness

Saturday 18th September
Left Sheerness 1.30pm for Berehaven with Formidable and Albermarle.

Sunday 19th Sept
Joined flagship HMS Prince of Wales at 5am and proceeded to Berehaven.
Kept afternoon watch.

Monday 20th Sept
Passed Lands End at 8am and carried out Control exercises in course of morning but these had to be abandoned because of fog.
Kept afternoon watch.

Tuesday 21st Sept

Arrived Berehaven 7am.
Midshipmen left for *training ship HMS Doris*.

Wednesday 22nd Sept
Ashore at Furious pier. Cycled to Castletown and back to Adricole. Returned from Furious Pier 7pm.

Thursday 23rd September
Remained on board printing and toning photographs.

Fri Sat24th 25th- September
Golf, tennis, walking,

Sunday 26th September
Church, climbed hill abreast of Furious Pier, heavy mist at top.

Monday 27th September
Ashore walk to Adricole 7 miles from Furious Pier with Whish. Tea at Adricole & returned on board 7pm.

Tuesday 28th September
Pulling Regatta for boats of Atlantic Fleet.
Results
1. Prince of Wales.
2. Queen.
3. Implacable.
4. Albermarle.
5. Formidable.
Implacable's gig disqualified after winning several races owing to being a Maltese boat. Ashore in afternoon and played golf with Ward.

Wednesday 29th Septenber
Another Pulling Regatta for Atlantic Fleet. Ashore in afternoon for tennis with Robertson, Laffen and Robinson.

Thursday 30th Septenber
Sailing Regatta for Atlantic fleet

Friday 1st October 1909
Winter routine begins. No cap covers. Ashore Bere Island Golf.

Saturday 2nd October
Ashore Furious Pier. Walk to Castletown.

Sunday 3rd October
Church on board.
Away in skiff fishing at night.
Pollock 12 lbs 5 55S
32 lbs altogether.

Monday 4th October
Night Defence Practice in Bantry Bay. Layed out targets in day & practice at night with Search Lights. 8 in number. Sweeping.

Tuesday 5th October
Night Defence Practice in Bantry Bay

Wednesday 6th October
Night Defence Practice in Bantry Bay. Towed target for HMS Prince of Wales.

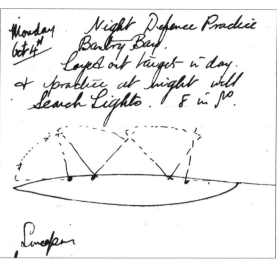

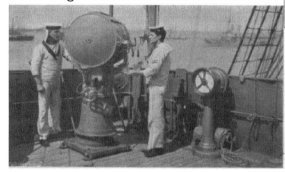

RAMC. Night Defence Practice

Thursday 7th October
Ashore at Bere Island for Golf. Night Defence Practice.

Friday 8th October
Ashore at Bere Island for Golf with Capt RAMC. Night Defence Practice

Saturday 9th October
Ashore at Castletown. Wet through so returned to hotel & got dry. Left Berehaven for Blacksod Bay 5pm.

Sunday 10th October
Outside Blacksod Bay. Passed wrecked TBD HMS Lee on rocks *after collision on 5th October*[39]. Ashore with Charig and Robinson. Tea at Hotel. Ship left for Swilly 6pm.

---

[39] Torpedo Boat Destroyers were only 350 tons but top speed 30 knots which was twice that of capital ships. Their distinctive turtlebacks were intended to clear water from the bows but instead dug the bow in, resulting in a wet conning position and only able to reach top speed in perfect conditions.

Monday 11th October
Arrived Loch Swilly 9.30am Ashore at Buncrana. Golf with Robertson.

Tuesday 12th October
Ashore at Buncrana. Walking.

Wednesday 13th October
Ashore at Buncrana Golf 3 rounds

Thursday 14th October
Rigged Target for Night Defence Practice and Implacable went outside in morning, but had to return to harbour at 4pm on account of bad weather and Practice abandoned.

Friday 15th October
Practice abandoned altogether.

Saturday 16th October
Ashore at Buncrana. Golf.

Sunday 17th Oct
Left Loch Swilly 4.30pm for Berehaven in company with Prince of Wales, Albemarle, Formidable and Venus.

Monday 18th Oct
Firing 6" Lyddite shells at 'Bills Rock' off Blacksod Bay to compare with normal Cordite explosive. 6 Rounds per ship.
Lydite as used for bursting charges in shells consisits of Picric Acid fused at 130° to 140°C and cast into the shell, bearing a central core for the exploding priming.

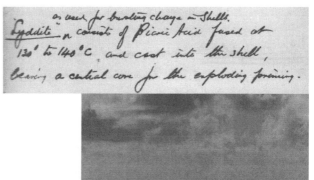

Tues 19th October
Arrived Berehaven 12 noon.

Wed 20th October
Ashore at Bere Island. Walking.

Thurs 21st October
Golf at Bere Island. 2 rounds 64+53 less 24 handicap= 93Nett.

Friday 22nd October
Ashore at Bere Island Golf with Robertson and Robins.

Saturday 23rd October
Ashore at Bere Island. Golf. Tea at Club

Sunday 24th October
Church on board Fore Mess Deck.

Monday 25th October.
Admirals Inspection.
Gunnery and Torpedo.
Admiral Prince Louis of Battenburg came on board at 8am.
Proceeded to sea 8.15am. HMS Prince of Wales towed target. Plotting etc. Fired 2 torpedoes, and returned to harbour 1pm.

During the Inspection, the fore top was supposed to be shot away and ship (guns) controlled by Main Top and finally guns put under local control[40].

In firing the torpedoes, 1 torpedo was fired at a target & the tube recharged & fired as quickly as possible.

Tuesday 26th October
Barometer went down very rapidly and owing to the storm, no boats were available.

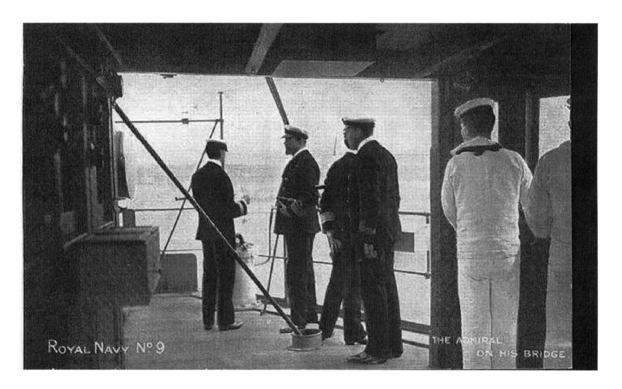

---
[40] Local control means individual gun crews aiming and firing independently without overall coordination.

Wednesday 27th October
Left harbour at 8am to tow target for HMS Formidable's Admiral's Inspection.
H.M.S. Formidable fired a shell which came over quarter deck of Implacable.
Length of the towing hawser had to be increased for the afternoon's Practice.
Towing target in afternoon for HMS Prince of Wales' Inspection.
Returned to harbour 4pm.

Thursday 28th October
Ashore at Bere Island. Golf with Robertson.

Friday 29th October
Court Martial on board HMS Implacable of Captain of Torpedo Boat Destroyer Lee which *was* wrecked in Blacksod Bay.
Ashore at Bere Island playing Golf handicap with Mortimer. Beat M. 5 up 4 to play.

Saturday 30th October
Ashore Bere Island. Golf Bourne. Won 6 & 5 with 9 handicap.

Sunday 31st October
Collier alongside for tomorrow's coaling 4.30pm.

Monday 1st Nov 1909
Coaled ship 6am–10am 400 tons.
Ashore at Bere Island Golf.

Colliers alongside HMS Russell

Tuesday 2nd November
Ashore Bere Island. Golf. Lost to Chaplain 1 hole. Handicap 14.

Wednesday 3rd November
Ashore Bere Island. Golf. Robinson.

Thursday 4th November
Ashore Bere Island. Golf Engineer Commander.
HMS Venerable joined up with Fleet.

Friday 5th November
Torpedo trials in Bantry Bay.

Saturday 6th November
In morning Admiral Inspected all the ships of the Atlantic Fleet in "Getting in and out Net Defences [41]" also "Boats, Anchors etc".

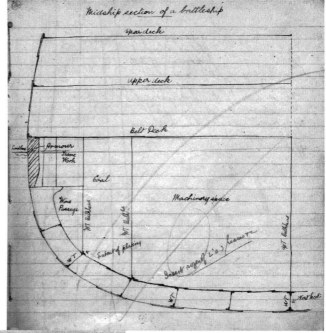

---

[41] Torpedo Protection was by a belt of armour above and below the waterline as in Albert's shipwright's mid-ship section sketch. Nets could also be deployed as tested "Out net Defences". Although made of steel, the nets could only be used when stationary and became obsolete when torpedoes with net cutters at the nose were invented.

Sunday 7th November
Left Berehaven 8.30am for Kingstown[42] in company with Prince of Wales, Venerable and Formidable.
Passed Lusitania at 12noon bound West.

Monday 8th November
Arrived at Kingstown 10am, went ashore and tram to Dublin, Nelson's Pillar. Returned on board 7pm.

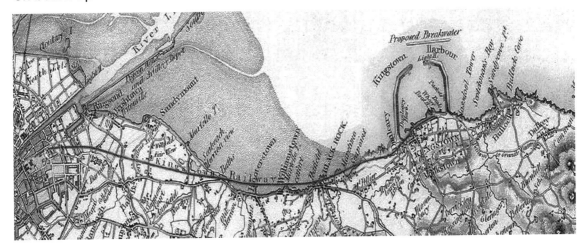

Tuesday 9th November
Remained on board

---
[42] Now Dún Laoghaire, Dunleary

Wednesday 10th November
Ashore Kingstown 11am. Train to Westland Row, Dublin.
Tram to Phoenix Park and Zoological Gardens and bookshops around town.

Thursday 11th November
Ashore Kingstown and walked to Monkstown with Bourne for Rugby football match. Fleet 19 points v Monkstown 6 points. Walk back to ship 7pm.

Friday 12th November
Left Kingstown for Sheerness 6am.
8 hours full power trial 7.30am-3.30pm.
16 hours 3/5th power trial till 6.30am on Saturday.
Working Indicator diagrams.

Saturday 13th November
Kept Afternoon watch. Changed over from Starboard to Port. Steering engines in afternoon. 4 men in steering compartment working clutches and E.R.A. (Engine Room Artificer) placing pins in Engine Room.
Arrived off the Nore about 9.30pm and anchored.

Sunday 14th November
Went into Harbour and moored to buoy off Victoria Pier 2pm. Very wet.

Monday 15th November
Ashore at Sheerness called at Tom Adams to tea, returned on board 7pm.

Tuesday 16th November
Ashore at Sheerness called Uncle Tom to tea

Wednesday 17th November
Ashore at Sheerness

Thursday 18th November
Left Sheerness 4.30pm on leave.
Left Paddington 9.15pm for Pembroke Dock.

Chapter 8   To Gibraltar

Monday 6th Dec 1909
Arrived back on board 10pm

Tuesday 7th December
Coaled ship 900 tons.

Wednesday 8th Dec
Left Sheerness 9.30am for Dover. Arrived 3pm. Ashore.

Thursday 9th Dec
Admirals Inspection in Section A.
Left Dover 3.30pm bound for Gibraltar. 11 knots at 66 revs.

Friday 10th Dec
Good passage down Channel, passing Ushant about 7 pm.

Saturday 11th Dec
Crossed Bay of Biscay.

Sunday 12th Dec
Steaming down coast of Portugal.

Monday 13th Dec
Passed Lisbon about 12 noon.

Tuesday 14th Dec

Arrived Gibraltar 1.30pm and moored No. 4 buoy.
Ashore 2.30 to meet *brother* George preparing Torpedo boat on slip in dockyard.
Called to see Emily

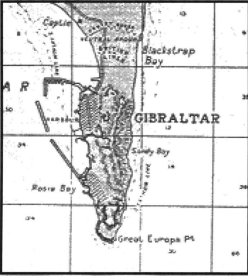

(George's wife) D9H10 Town Range[43]

Wednesday 15th Dec
Lecture on History of Ship Building.
Ashore 3.45 George.

Saturday 18th Dec
Ship went alongside dockyard wall for refit.

---

[43] D9H10 possibly their address in barrack accommodation or slipway number or Torpedo boat number.

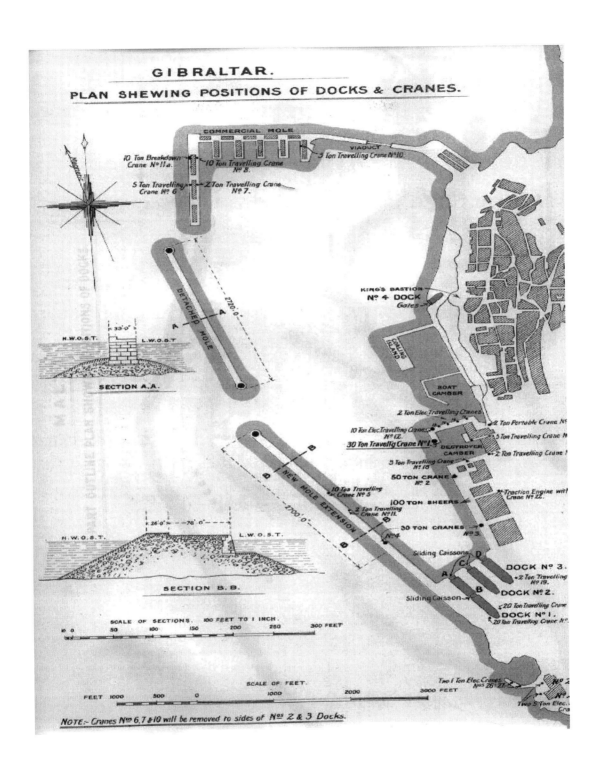

Ashore and inspected water catchments on the Rock with George and Emily.

Sunday 19th Dec
Wesley Chapel morning and evening. Tea at George.

Monday 20th December
Cruiser HMS Donegal collided with S.S. Malaga in the harbour and had to dock. Called at George.

Tuesday 21st December
Tennis.
Walk round Alameda Gardens.

Wednesday 22nd December Thursday 23rd December
Tennis

Friday 24th December
Walk up rock to Signal Station with George. Spent evening with George.

Saturday 25th Dec Christmas day. Service at Wesley Chapel. Spent day with George. Walk round rock.

Sunday 26th Dec
Wesley Chapel morning and evening.

Monday 27th December
Remained on board.

Tuesday 28th December
Class meeting at Wesley Chapel. Leader Rev Brown.

Wednesday 29th December
Remained on board.

Thursday 30th December
Walk up rock past Wireless Station and up Windmill Hill.

Friday 31st December
Dinner with Mr & Mrs F Vaughan.
Watchnight Service at Cathedral.
Turned New Year in at George.

Saturday 1st January 1910
Walk with George and Emily to Cemetry and Racecourse.

Sunday 2nd January
Service at Wesley Chapel. Communion Service. Tea at George.

Monday 3rd January,
Remained on board.

Tuesday 4th January
Remained on board.

Wednesday 5th January
George.

Thursday 6th January

Implacable went into Dock No.3 for refit at 8am. Collision lines left on and diver sent down to see if blocks had tripped. 8.50am. Caisson closed and pumping began at 21000 tons per hour with 7 ft pumped out in 20 minutes.
9.20am. set up breast shores[44].
11am. Water over sill 1feet. 2 tons of shores. Part of ribband shores
12am. 3tons of shores up. Water over sill.
The ribband shores were now inconvenient in rigging stage.

Friday 7th January

George.
Took Photo of Implacable in dry dock.

Assistance in dry dock.

Saturday 8th January

French fleet of 6 Battleships arrived and moored to Buoys.
Crossed to Algeciras by 2.50 boat with George. Cathedral and Bullrigs church. Returned to Gibraltar 6pm.

Sunday 9th January

Service at Wesley Chapel morning. Tea at George. Service Wesley Chapel evening with George and Emily.

10th-14th January *blank*

---

[44] Timber struts to keep ship upright in dock as seen in Albert's photo.

Saturday 15th January
First day of elections at home. Prayer meeting at Wesley.

Sunday 16th January
Wesley Chapel. Lunch Mediterranean Club with N.S.O. Chapel in evening. Missionary meeting.

Monday 17th January
Moorish Castle and Prison. George and Emily cut pieces from scaffold on which a Calvo murderer was hung. Missionary meeting at Wesley.

Tuesday 18th January
Walk round Catalan Bay with George. Oil tanks, Water catchments etc. with George.

Wednesday 19th January
Walk to Governors Cottage and Monkey's Cave (Europa) with George.
Election Pembroke Dock result Liberal majority 705.

Thursday 20th January
Walk round Eastern Beach with George.

Friday 21st January
Tennis with Charig, Jones & Robinson. Dock partly flooded 3.30pm preparatory to coming out of dock.
3 tons of shores left under ship. When ship was floated off.
Election at Pembroke County. Radical majority

Saturday 22nd January
Undocked and proceeded to buoy.
Ashore and walked to Queen of Spains Chair, Spain. Inscription "Weyler. Primer ministro de la Guerra que me visitola patria aggradacida. San Rogue 22 September 1901"[45]

---

[45]Translation-"Prime minister of the War visited the grateful homeland. San Rogue 22 September 1901"

Sunday 23rd January
Church on board. Wesley Chapel at night with Emily.

Monday 24th January
Remained on board.
Prepared to coal from lighters.

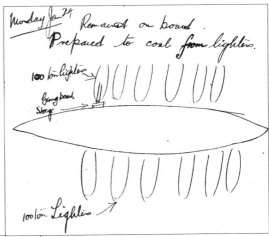

Tuesday 25th January
Coal ship from lighters
1400 tons 7.50am to 7pm.
Averaged 144.5 tons per hour.

Worked by baskets carried by sailors.

Wednesday 26th January
Ashore 4pm

Thursday 27th January
Ashore 4pm

Friday 28th January
Ashore.

Saturday 29th January
Ashore in evening.
Met Mr Lewis.

Sunday 30th January
Wesley Chapel in morning. George to tea. Cathedral in evening.
Walk round Gibraltar South after service.

Monday 31st January 1909

Called Mr Lewis in evening with George.

Tuesday 1st February
Called Mr Lewis in evening with George and Emily.

Wednesday 2nd February
Anniversary of commissioning of HMS Implacable. Ashore 4.30 shopping with George and Emily. Called on Chief Constructor at Dockyard.

Thursday 3rd February
French Cruiser 'Chateau Renault' lost her rudder after a grounding on African coast and went into No. 1 Dock for repair.

Deck cut away to admit firing of forward gun.

Trim of ship about 5 feet.
Boat davits built up *with* lightened plates and angle bars.

Basin Trial[46] of Implacable.
IHP at 30 revs=833. Ship secured by hawsers alongside.

At 30 revs when steaming 420 IHP.

---

[46] Stationary Steam Trial. Albert makes no comment on the result showing twice the power compared with the ship moving.

Friday 4th February

Had photograph taken at Rock Studio in dress Uniform. Dinner with Mr & Mrs Vaughan.

Saturday 5th February

Walked round Eastern beach with George and Emily. Received appointment to the Admiralty this app^t dated 2nd February via Payne (overseeing).

Sunday 6th February

Cathedral in morning. Received Photographs.
Wesley in evening.

Monday 7th February

Steam trials 9am-12 noon. Ashore in afternoon. Called Mr Evans
Aboard 11pm.

Tuesday 8th February

Left Gibraltar 7am for England at 60 revs.

Wednesday 9th February

Dined Warrant Officers in Ward Room.

Thursday 10th February
Entered Bay of Biscay 8pm.

Friday 11th February
Crossing the Bay had very rough passage. Waves *same as* period of ship.

Saturday 12th February
Arrived off Ushant. Sea still running very high.

Sunday 13th February
Fair passage up English Channel.
Passed Dover in afternoon where several ships of Atlantic Fleet in Harbour.
Arrived at Sheerness at 6pm. Landed & relieved crew of HMS Dwarf, a 900ton gunboat.

End of voyage and my sea time.

Monday 14th February
Left ship for London.
Saw Director of Naval Construction and commenced duties at Admiralty[47].
Started drawings of Colonial Class Cruisers.

---

[47] Firstly in Admiralty Arch, London, later in Admiralty offices overlooking Horse-guards Parade and relocated to Bath, Somerset during WW2.

Addendum

Albert 'Ajax' Adams spent his whole career designing Cruisers including HMS Exeter, Norfolk, Achilles, Sydney and Ajax.
Albert's nickname was derived from his HMS Ajax shown below.

Launched 1st March 1934, his only surviving son, Robert, was born 4th March 1934. A close-up of Albert in top hat, top left, is taken from the larger photo opposite at the 1935 launch of his HMAS Sydney.

He did not design HMS Belfast ^{below} which is now preserved in the river Thames. When this ship was torpedoed in WWII, he designed a quick repair by adding a foot thick plating around the hull. This scarcely affected the speed and enabled the ship to rejoin

the war effort. He was still alive when she was converted into a floating tourist attraction but never wished to visit her as a museum.